The Quick Guide to Robotics and Artificial Intelligence

SURVIVING THE AUTOMATION REVOLUTION FOR BEGINNERS

ALEX NKENCHOR UWAJEH

LEGAL DISCLAIMERS

All contents copyright © 2017 by **Alex Nkenchor Uwajeh**. All rights reserved. No part of this document or accompanying files may be reproduced or transmitted in any form, electronic or otherwise, by any means without the prior written permission of the publisher.

This book is presented to you for informational purposes only and is not a substitution for any professional advice. The contents herein are based on the views and opinions of the author and all associated contributors.

While every effort has been made by the author and all associated contributors to present accurate and up to date information within this document, it is apparent technologies rapidly change. Therefore, the author and all associated contributors reserve the right to update the contents and information provided herein as these changes progress. The author and/or all associated contributors take no responsibility for any errors or omissions if such discrepancies exist within this document.

The author and all other contributors accept no responsibility for any consequential actions taken, whether monetary, legal, or otherwise, by any and all readers of the materials provided. It is the reader's sole responsibility to seek professional advice before taking any action on their part.

Reader's results will vary based on their skill level and individual perception of the contents herein, and thus no guarantees, monetarily or otherwise, can be made accurately, therefore, no guarantees are made.

THE QUICK GUIDE TO ROBOTICS AND ARTIFICIAL INTELLIGENCE

Table of Contents

Introduction 6

Understanding Robotics, Artificial Intelligence and Automation 10

Robotics and Industries 15
    Jobs That May Vanish in Future Years 16
    Back-office jobs 17
    Consumer services and Taxis 17
    Construction 18
    Postal Service 19
    Accounting 20
    Legal 20
    Medical Diagnosis and Health Care 21
    Banking 22
    Stock Brokers 23
    Journalists 25
    Movie Stars 25
    Translators 26
    Receptionists, Secretaries and Personal Assistants 27
    Hospitality 28

History Can Reveal the Future 29

The Robot Revolution and Investing in
Robotics 32
    Investing in Robotics 33

Effects of Technologies and Automation 36
    Impact on Incomes and Benefits 38
    Impact on School Curriculum 40

Surviving the Automation Revolution 43
    Preparing for the Automation Revolution. 46

Artificial Intelligence Revolution Begins 49
    The Road to Super-Intelligence 50
    AI In Everyday Life 52
    AI in the Future 53
    The Element of Innovation 56

Conclusion 60

Introduction

"We are on the edge of change comparable to the rise of human life on Earth." ~ Vernor Vinge

When many people think of artificial intelligence (AI), they cast their minds to many big-budget science fiction movies that show AI computers becoming self-aware, taking over the world and dominating humanity. Ask those same people what they think of robotics and they immediately associate them with automation and taking over everyone's jobs.

What you may not realize is that robotics and AI technology have already become a part of mainstream live as we know it. Many of the technologies we take for granted today have been around in some form or other for more years than most people realize.

AI is not a new concept. In fact, it's been around since 1956, when a group of people had the idea to build a machine that was as intelligent as a human. Using a programmable digital computer created back in the 1940s, the same group of scientists began working on the possibility of building an artificially intelligent machine, or computerized brain that was able to learn.

Technology has advanced significantly since inception, allowing developers and researchers to integrate AI programming and robotics into things you probably already use in your daily life.

In fact, robotic automation has become so mainstream in many things we do that their presence is already taken for granted.

Think about the automatic doors that slide open to allow you to enter the mall without you touching a thing. The soda bottle you drink from was filled, labeled, sealed and packed ready for shipping to the 7-11 using robotic automation. The plastic bottle itself was most likely also created using robotic automation in a large plastics injection molding company somewhere.

While many people in the general population have no real idea what to think about artificial intelligence, anyone who works within the industry believes it's able to bring about the biggest change in humankind's history.

When you look back at some of humankind's most progressive changes and events, a few things really stand out: the intentional use of fire, the invention of agriculture, the industrial revolution, the invention of computer, and the advent of the internet.

Those significant events changed the way people live, think, and work and created history. By comparison, advanced artificial intelligence is likely to leave those historic events in the dust.

No matter what your personal thoughts on robotics, automation or AI might be, the revolution is coming. Those who are most likely to thrive and prosper in the AI era are those who embrace the technology and prepare for the revolution ahead.

Understanding Robotics, Artificial Intelligence and Automation

If you turn on a light switch or plug an appliance into an electrical socket, you've learned to trust that electricity makes those things work. In order to benefit from electricity, it's not necessary for you to understand how it operates behind the scenes. Only that it allows you to achieve the outcome you hoped to achieve by its use.

The same thing is true with robotics, automation and artificial intelligence.

In order to understand the impact technology will have in our future, it's not always necessary to completely understand its inner workings or how those things operate on an internal scale. You simply need to understand how it has the potential to impact you and your life – and how you might take advantage of it going into the future.

If you're like many people, it's likely you'll see artificial intelligence (AI) in terms of a self-aware science-fiction character hell-bent on dominating and wiping out humankind. They immediately think about humanoid robots, like the ones portrayed in Hollywood films like Terminator and automatically assume that AI is the same thing as robotics.

The reality is that many people simply don't understand what AI really is and what it's capable of achieving.

Yet it's likely you already use AI in your everyday life, even if you don't realize it. The calculator app on your smartphone uses

AI programming. Automated self-driving unmanned vehicles use AI programming. If you use an iPhone, the software and programming behind Siri is all AI.

Many of the amazing technological things we use today are taken completely for granted. Yet they originated from simple ideas and a determination to bring those ideas to life using technology that was already available in some form and may only have needed the right implementation to achieve significant results.

It's common for many people to assume that robotics, automation and artificial intelligence are all relatively new inventions, created within the last decade or so. While it's true that many of the world's more astonishing inventions have been circulated to a broader global audience than ever before in recent years, the technology behind many of today's robotics and automotive processes have been around for many decades.

The original ideas and processes may have been designed and developed years ago. However, it's only with today's super-powerful computer processes and wide availability of technological tools that many of those original ideas can now be expanded upon to create the incredible leaps of forward technological momentum consumers see at an ever-increasing pace today.

The field of artificial intelligence often receives little to no credit for its contribution to many past successes. Some of the greatest innovations that resulted from AI tend to be reduced in significance once they are filtered into general acceptance.

Once anything becomes useful and is used commonly enough, it's no longer labeled as being 'AI'. It simply becomes a computer algorithm, programmed by people to achieve a desired outcome.

The same is true with the field of robotics. The automated machines already in use by

1900 would have seemed fantastical to someone back then, yet they now seem so commonplace to today's consumers that it no longer seems innovative at all.

Robotics and Industries

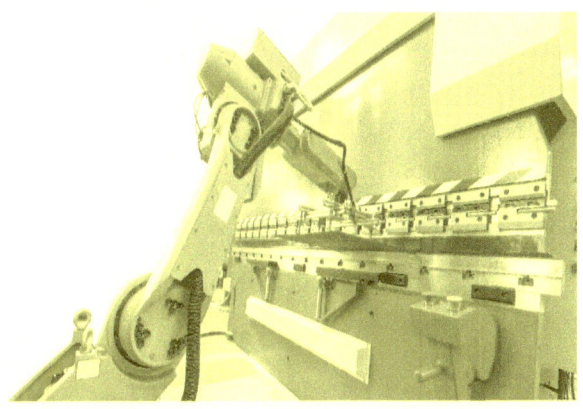

Robotics and automation have the potential to improve productivity and profitability for many industries around the world. However, there are plenty of people who only see robotics as a source of impending doom for hundreds of thousands of jobs.

Popular media, in particular, is fond of painting an image of a bleak, dystopian future where intelligent computers control everything. There are thousands of reports being published in major newspapers or reported in newscasts around the world every month about how many people may find their jobs have been replaced by robotic automation or superseded by AI software programming.

When the subject of robotic automation is broached, most people instantly associate it with the loss of many manual labor jobs. Manufacturing and manual laboring jobs are typically the first casualties people think about when they consider the effect of robotic automation.

However, there are plenty of other skilled jobs and even entire industries that may disappear over the coming years as technology continues to advance.

Jobs That May Vanish in Future Years

While many people worry about the future of manual jobs, they don't often consider the impact automation could have on a vast cross-section of other industries and jobs. In fact, AI technology has the potential to replace or completely transform a number of professions and make many of today's jobs redundant.

Back-office jobs

Business owners are able to take advantage of AI programming and cognitive technology to automate many back-office processes. Automating once-costly manual tasks promises to help improve productivity and boost profitability, but at the same time eliminates many back-office jobs and manual functions, displacing a large number of employees.

Consumer services and Taxis

A report released by Forrester predicts that, in the nearest future automation, whether in the form of mechanized robots or as artificially intelligent software programs, will eliminate up to 6% of jobs in industries that include customer service, logistics, trucking, consumer services and taxi services.

Self-driving vehicles are already being tested that could easily be programmed to become the new norm for taxi services. In fact, Singapore is already testing a dozen self-driving taxis to determine viability and safety.

Autonomous vehicles are eagerly anticipated by many ride-share companies, including Uber. The ride-sharing company's CEO understands that rides would be cheaper and profits would be larger if passengers didn't need to pay for the driver's time.

Construction

Manual labor jobs are also under threat. Robotics and automation will soon be able to replace some construction workers. Technological advances in 3D printing integrated with concrete pouring controls allow entire buildings to be constructed faster without the need for manual labor.

Technology and automation are even apparent at the grocery store. More people choose to use the self-checkout option to pay for their items, which reduces the need for store owners to hire so many cashiers.

Postal Service

People working at the postal service are quickly being replaced as more people become comfortable receiving with electronic bills and online greeting cards instead of paper communications.

The emergence of driverless vehicles may also impact jobs within the postal service, as more autonomous drones and unmanned vehicles become more commonplace.

If the current paperless trend continues, the postal service may need to think about transitioning to bigger box deliveries to cater for the increased number of deliveries for people purchasing their goods online.

Accounting

Research conducted by the University of Oxford predicts a 95% probability of artificially intelligent software programs replacing accountants in the near future. Advances in accountancy software could see a large number of accountants scale back the number of employees working manually through client files.

Legal

It's also estimated that the legal field could also suffer job losses as automated software programs begin to replace a number of skills and competencies. Associates and paralegals may no longer be required to spend hundreds of hours researching obscure legal references when AI programs can locate the required information in a fraction of the time.

Medical Diagnosis and Health Care

Many companies are already using AI programs to aid clinicians in the field of medical diagnosis. Current methods for detecting medical conditions such as cancer can be inaccurate and expensive. They're also often uncomfortable for the patient.

Deep learning-based diagnostic models help to improve the accuracy rates for discovering and diagnosing serious health conditions using less invasive methods overall. Integrating remote monitoring sensors and machine-to-machine communications can automate many health care and diagnostic processes, effectively removing the need for humans to complete the same tasks.

Remote sensors and monitors can also be used to remove the need for nurses or health care staff in some situations. A doctor has the capacity to monitor a patient's blood pressure, heart rate, and blood oxygen levels using remote sensors that transmit reports directly to the doctor, who is then able to adjust medications or treatments accordingly.

Banking

Banking jobs will also be impacted. The vast majority of people take ATMs for granted these days, but when they were first introduced a large number of bank tellers and customer service representatives lost their jobs.

As online banking programs become more sophisticated and integration with smart phone apps continues to improve, the need to maintain and staff physical bank branches is dramatically reduced. Customers can open new accounts online and submit loan applications with a few clicks of a mouse button.

There are already banks in Australia experimenting with unstaffed automatic banking branches in some areas, taking advantage of automated teller machines to accept deposits of cash and checks at any time of the day or night. There's no need to wait for branch opening hours, which is especially convenient for business banking clients.

Customers simply swipe a keycard to verify that they are customers of that bank in order to enter the secure, automated branch. The doors automatically close behind them and security cameras monitor every person within the branch for added safety.

Once inside the automated branch, the customer is able to safely deposit coins, notes or checks into the machine, or withdraw cash, or transfer funds between accounts without the need for bank tellers.

Stock Brokers

Gone are the days of hundreds of frenzied stock brokers standing on the trading floor of the stock exchange. Computerized algorithms are already able to take the place of human transactions.

Traders have been able to buy and sell stocks using a computer trading platform for several years already, reducing the need to pick up the phone and call a stock broker to place the same order on their behalf. The computer program is able to match the trade automatically and facilitate the transaction without human intervention.

Arbitrage traders are also able to benefit from improvements in computer algorithms. The objective for many day-traders is to buy stocks low and sell them at a profit when stock prices increase in price sufficiently.

Rather than sitting in front of a computer analyzing various stocks for likely trading opportunities and then waiting for stock prices to change, it's possible to set the computerized trading platform to complete these tasks automatically. The program is able to watch a significantly larger selection of stocks for emerging pricing opportunities, place the order to buy the stock, and automatically close out the trading position once a profit has been realized.

Journalists

Advances in artificial intelligence software may also completely replace the jobs of journalists and writers. Publishers may be able to program intelligent software to write the articles and reports needed on demand without relying on a human writer at all.

Movie Stars

Perhaps one of the most overlooked industries that could see major job losses and enhancements in coming years is the movie industry. Actors are already being replaced with animation and CGI-enhanced graphics. Even deceased actors can be resurrected using CGI techniques. If technology continues to improve, there may be fewer new jobs for actors, as many existing stars have the potential to continue appearing on screens via CGI programming for decades to come.

Translators

Professional translators and interpreters were once in high demand right across the world across a number of industries. Tour guides are perhaps the most commonly associated profession for many translators.

However, translators were also in demand at conferences or speeches. Interpreters are also needed in the legal profession and in courtrooms to translate testimony given by someone who speaks a different native language. International business transactions are often facilitated by translators. Written-word translators were also in demand to translate manuscripts and other written information into different languages.

Consistent advances in translation software programs have made it easier for people to translate the things they need using a computer or smartphone, removing the need for a human translator in many instances.

Automated translation apps installed on smartphones allow tourists to understand what locals are saying in real time. Software programs can translate large sections of written text with increasing accuracy.

Receptionists, Secretaries and Personal Assistants

The human element of answering phones, responding to emails, or booking appointments in a diary or calendar is quickly becoming redundant.

In the case of personal assistant work, a virtual assistant programmed for intelligent response is no longer considered a futuristic vision. It's already a current reality.

There are already artificial intelligence programs being used as virtual assistants across some industries. The program is able to perform many of the same tasks as a human assistant, with the ability to read emails, discern the intent of the message, and create a relevant response. The program is able to book appointments and meetings.

Hospitality

Advances in robotic programming for humanoid robots have improved to such a degree to replace human workers in some hospitality jobs.

Humanoid robots are already being used in a number of hospitality jobs to reduce the need for customers to interact with human wait-staff. Multiple fast-food restaurants have already installed tablet PCs that allow customers to place their order and pay directly without talking to service staff or cashiers. The kitchen receives the order directly from the tablet and prepares the food.

Robots have already been introduced into a Japanese hotel to perform basic hospitality tasks. The robots have been programmed to take reservations, check guests in and out of the hotel, and escort them to their rooms.

The robots are also able to speak several languages, which is ideal for international guests. Once guests have arrived at the room, they are able to address the robot with voice commands to perform basic functions, such as adjusting the room temperature, alter lighting, or respond to simple questions about local weather conditions.

History Can Reveal the Future

While a large number of jobs as we know them today are likely to become defunct or be completely replaced by robotics or AI programming, it's important to note that new jobs will continue to be created as the need arises.

History shows us that a large number of today's jobs may become redundant as technology continues to change the way we do things, but a range of new jobs will emerge to replace them.

A report released by Deloitte looks to historical data and highlights the sheer number of jobs that were destroyed or eliminated during the first industrial era. Steam-powered machines increased production for a large number of manufacturing industries, which saw many manual laborers lose their jobs.

Many people opposed the impending Industrial Revolution in the late 1700s and early 1800s, afraid that machines would replace their jobs as technology continued to improve. Huge numbers of skilled laborers and craft workers across many industries lost their jobs, including those within textile manufacturing and weaving, metallurgy, glass making, paper making, agriculture, mining, transportation, gas manufacturing, chemical production, and many others.

Yet the jobs that vanished during that time were replaced with newly-created jobs that supported the flourishing new industries throughout that era.

The same thing occurred when personal computers became commonplace. Many people predicted that a large number of jobs would be wiped out and replaced by computer processes.

While plenty of jobs vanished or changed completely, a vast range of new jobs were created as a result. As more people embraced personal computers in business and home environments, the need for skilled staff to repair, update or maintain those machines and networks increased.

As the number of jobs being replaced by automated processes increases, it only follows that a vast range of new occupations will be created in their place.

The Robot Revolution and Investing in Robotics

According to a study conducted by RBC Global Asset Management, the cost of implementing industrial robots and automation has decreased significantly in recent years. As a result, the number of industrial robots used around the world has increased dramatically.

In 2013, there were around 1.2 million industrial robots in use globally. That number has continued to increase and is still growing.

The U.S. Bureau of Labor Statistics (BLS) released projections regarding future employment. The BLS predicts that approximately 15.6 million new jobs will be created within the next 10 years. The statistics take into account positions that may become obsolete or those that may be less in demand as technology begins to replace human workers.

There will be fewer people performing repetitive manual jobs in future, but new jobs managing data networks, analyzing big data or mining information will be created. No matter how many manual tasks are automated by machines, humans will still be needed to manage the digital world.

Investing in Robotics

Anyone with money to invest is able to begin investing in robotics, whether directly or indirectly.

Direct investment simply means putting up the money needed to pay for research and development of new robotic technology. In many cases, the military and governmental agencies are perhaps the largest sources of direct investment in the field of robotics.

The objective is to discover cost-effective new ways to use robotics to automate various processes that were once reliant on humans to complete them. Government agencies and military aim to implement the automated robotic processes that offer the opportunity to improve productivity and increase accuracy for each process being performed. Unfortunately, the same processes also eliminate the need for humans to complete those tasks, making many jobs redundant as a result.

Many large corporations, small business owners and private enthusiasts are also willing to invest heavily in research and development for innovative new ideas using robotic technology. The objective of investing in robotics for corporations and businesses is to potentially boost productivity levels and decrease operating costs, which ultimately boost profits.

Indirect investment is perhaps the easiest way for anyone to invest in robotics. Many tech companies are publicly listed on the stock market, so an investor only needs to place an order to buy stocks within their chosen company.

As a shareholder of the company, there are two potential ways to profit from your investment. The first option is capital growth. If the price of the company's stock increases, your own shares also increase in value, representing capital growth. The second option is through dividend income. Many companies pay investors a share of any profits made each year in the form of dividends.

Effects of Technologies and Automation

The full impact of emerging technologies and automation is difficult to determine, simply because we're still at the beginning stages of the technology revolution. A number of studies exist that explore some of the possible social, financial, and employment impacts, but the full extent of the true impact automation and emerging technologies will have is yet to be fully realized.

Many existing companies and organizations have already achieved a large economic scale without the need for a large workforce. Companies such as Google and

Facebook have built enormous company values and revenues using a comparatively low number of employees. Both companies rely heavily on computerized algorithms to achieve intended outcomes and user experiences.

The types of automated tools available today are no longer considered as the cutting edge of new technology. Companies and organizations all around the world are already taking advantage of robotics, artificial intelligence, computerized algorithms, 3-D printing technology, and unmanned vehicles, each making significant changes on the way people think and perform.

As automated tools become more commonplace, they create an ever-changing workforce that has the power to affect different demographic groups in a number of different ways.

Impact on Incomes and Benefits

If automation, robotics and AI continue to replace the need for workers at current rates, it's logical to predict that there will be fewer workers needed in future jobs. Technology may be destroying some jobs and creating new ones in their place, but it's becoming evident that more jobs are becoming obsolete than are being created.

The shortfall may translate into an economic situation where many people may no longer have the opportunity to earn enough discretionary income to survive financially with their current skills. The reality is that some people may be left behind as technology continues to progress.

There are also concerns about increases in income inequality. More people are likely to become effectively unemployable due to a lack in technological skills and the loss of many blue- and white-collar jobs. Some experts voice concerns about emerging technologies creating a permanent underclass of unemployable people

Many people also rely on their employment to access company-sponsored health care benefits, insurance, disability, and pensions.

As the number of workers needed in an economy based on emerging technologies is reduced, the ability for those people to gain the benefits and retirement plans their employers once provided is also greatly reduced. The impact will force workers to consider alternative options for earning income, gaining health care benefits, disability payments, and pensions.

Workers have the option of returning to study to improve education and job skills, increasing their chances of gaining full-time employment in other occupations. Another alternative displaced workers have is to follow an entrepreneurial path and create their own employment prospects as small business owners.

Impact on School Curriculum

The U.S. Department of Education provides some insight into how few students will be prepared for the automation revolution. Only a small percentage of American high school students are being trained in science, engineering, technology or math skills needed to thrive in a work environment focused strongly on technology, automation, and robotics.

When the statistics available from the U.S. Department of Education are broken down further, it becomes more apparent that racial minorities may face even greater challenges securing job opportunities than ever before.

Across many parts of the country, racial minorities already struggle with high unemployment rates due to lack of education and training. As technology continues to advance and many more unskilled jobs are automated, it will become more difficult for them to adapt to changes in the job market.

The current school curriculum is heavily geared towards teaching students a variety of skills applicable to jobs that may no longer exist in coming years. There is also the problem of more students choosing not to study in fields of science, math, engineering, or technology.

By the time many students leave school, it's likely they'll be underqualified to gain employment in anything other than unskilled labor or low-paying employment.

Amending and reforming school curriculums could help to ensure students are given the best possible chance of learning viable skills based on emerging technologies. Students may also benefit by having technology-based classes integrated into current curriculums to provide a basic introduction to potential new skills and to encourage increased participation in classes.

Traditional schooling systems are also quickly becoming outdated. More people are accessing open learning and online education courses to expand existing skill sets. The opportunity to broaden knowledge and improve education without attending a traditional school system is now greater than ever before.

Surviving the Automation Revolution

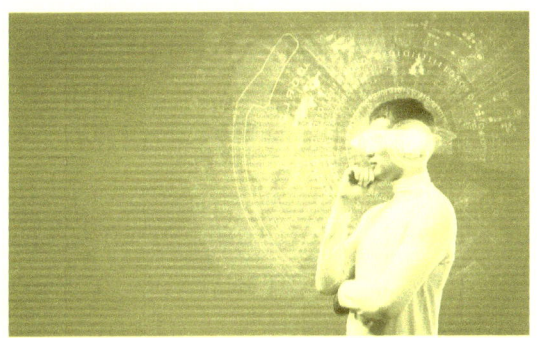

While a large number of today's jobs may eventually be taken over by robotic automation and AI programming, there are some professions and occupations that are much less likely to be impacted.

For example, nurses, doctors, psychologists, therapists, counselors, and social workers are some of the least likely types of occupations that may be taken over by automation, although many of the existing processes may change over time thanks to automation technology. These types involve empathy and require caring for others.

There are also some occupations that are more likely to survive the automation revolution by amending or expanding existing skills.

Economist Amy Rosen writes that the single most needed skill in a future jobs market will be an entrepreneurial mindset. Innovative thinkers who understand how to embrace robotics, automation and AI and use technology to their advantage have the ability to become founders of some of the more successful businesses of the future.

People with an entrepreneurial mindset are also likely to be more able to navigate the ever-changing innovation economy.

Innovation economy isn't a new term. In fact, the term was coined back in 1942 by Joseph Schumpeter. The basis of innovation economics is based on two fundamental elements: improving productivity through innovation; and that technological change is at the heart of economic growth.

Some real-life examples of successful industries that have been created and expanded by people with an entrepreneurial mindset include:

- Digital media technology
- Nanotechnology
- Biotechnology
- Information technology
- Precision engineering

Digital technology provides ordinary people with the best economic conditions in history to create and develop businesses that create their own consumer demand. Digital media businesses are prime examples of entrepreneurial people taking advantage of technology and creating new business models and employment opportunities that never could have existed in the past.

Likewise, some biotech companies take an entrepreneurial mindset and link it with the science sector to create new innovative activities and future economic growth.

Preparing for the Automation Revolution

In order to really prepare for the automation revolution, it might be helpful to look back at how previous generations survived the last major revolution: the industrial revolution.

Before the industrial revolution, many people relied on their entrepreneurial instincts to survive. They often ran 'mom-and-pop' stores, operated small business enterprises, provided in-demand services, toiled in agriculture or fishing industries, or worked as laborers in local small businesses. Many people lived in decentralized locations, as there was no real need for them to head into cities to work.

When steam-powered machines and automation replaced a large number of manual labor jobs, huge numbers of people were forced to move into cities to gain employment in factories. The result was massively centralized populations living and working wherever the income could be derived.

If you transpose the changes from the last major revolution to the future of the automation revolution, you see many things working in reverse. The need to live in or near a city in order to earn an income is rapidly reducing.

More people than ever before are working from home, thanks largely to the advent of the internet and the speed and power of personal home computers. Working from home reduces long commuting times and reduces costs associated with traveling to and from an external workplace.

An increasing number of people around the world are able to build wildly successful businesses using the internet and digital media technology, creating new employment opportunities in emerging businesses that aren't necessarily in centralized locations. The business's head office could be located anywhere in the world and employ workers from various locations who aren't expected to commute into the office to complete their jobs.

Those new business models are more likely to take advantage of AI programming, automation technology and digital media to improve productivity, market and promote new products or services on a global scale, attract new customers electronically, facilitate payment processing remotely, and deal with stock and order fulfilment with the click of a button.

The jobs of the future may rely heavily on innovation and creative thinkers. Jobs that require workers to deal directly with people aren't likely to change too much, although many of the processes currently used may become automated to some degree.

Artificial Intelligence Revolution Begins

When you really think about it, the AI revolution has already begun. It began decades ago.

Creative inventors and innovative thinkers are often responsible for some of the world's greatest technological inventions. It's possible the next big technological breakthrough could emerge from a basement or dorm room, designed and created by a home enthusiast or a researcher working on a completely unrelated field.

The Road to Super-Intelligence

Artificial intelligence is a broad concept. Many people just assume it means a self-thinking, self-aware, constantly learning computer.

Yet AI is a term that can be used to describe any computerized algorithm or process. In an effort to break down some of the differences between what's possible and what it can be used for, the industry has separated AI into three primary AI caliber categories.

These include:

- Artificial narrow intelligence (ANI): Artificial narrow intelligence, or sometimes called 'weak AI', is a specific type of AI that only specializes in one area. For example, the AI program that is able to beat a world chess champion in a game of chess is ANI, because that's all it does. It's incapable of achieving other objectives other than what it was programmed to do.

- <u>Artificial general intelligence (AGI)</u>: Artificial general intelligence, or sometimes called 'strong AI' or 'human-level AI' is the term used to describe a computer program that is capable of being as smart as a human across a range of processes. It's capable of performing any intellectual task a human being can achieve, including having the ability to reason, solve problems, understand complex ideas and thought processes, learn from experience, and think abstractly.

- <u>Artificial super-intelligence (ASI)</u>: Super intelligence is defined as being an intellect that is smarter than the best human brains across every field of knowledge and social skills. Artificial super intelligence computers can range from being only marginally smarter than humans to ones that are infinitely smarter in every aspect.

AI In Everyday Life

Many of the things we take for granted in our daily lives are already reliant on AI processes, whether you realize it or not. For example, the smartphone you carry around in your pocket is programmed with artificial narrow intelligence (ANI) systems. When you open your email inbox and check your spam filter, it's probably never occurred to you that the programming that separates out some emails from others uses ANI programming.

Modern cars use various ANI systems, from the computerized braking systems, to the inboard satellite navigation systems and fuel injection systems. More than half of the equity shares traded on the US stock market are completed using ANI programs to place the order and facilitate the transaction.

When you conduct a search on Google or Amazon, the results you receive arrive their via sophisticated ANI programming. Google Translate uses ANI systems that are incredibly good at completing one narrow-focused task.

When you consider some of the everyday applications for artificial narrow intelligence programming, they're not particularly scary or likely to take over the world and annihilate human existence as we know it.

AI in the Future

When applications go from ANI to artificial general intelligence (AGI), things become more interesting. AGI programming hasn't yet made the transition to commonplace processes and everyday life as yet, but many companies are investing heavily in research to make it happen.

The biggest leap for researchers and programmers is working on the right algorithms that allow a computer to actually think like a human. Various thought processes are difficult for humans, such as complicated math problems, language translation, or financial market strategies and a computer is able to complete the task easily.

However, give a computer a task that a human finds easy, such as vision, movement, motion, or perception, and the computer programming behind it all is intensely difficult.

When you boil it right down to basics, a computer is able to successfully do anything that requires humans to really think about achieving. Those complicated thought processes take very little effort for a computer program to solve.

But it struggles to do those simple things that humans are able to achieve without thinking. The simplest movements, perceptions and abstract ideas people have are insanely difficult for a computer program to achieve.

As computing power continues to increase at staggering rates, the potential for improving current advances in AGI grows exponentially. To many people, creating enough computer power to emulate a human brain still sounds like science fiction.

However, researchers have already been able to emulate the brain of a flatworm. It's true, the brain of a flatworm is nowhere near as advanced as a human brain. But as the computing power of computers continues to grow, the ability for researchers to transpose existing programs into emulating the brains of larger organisms also grows.

Before long, researchers will be emulating brains of fish, mice, rabbits, cats, dogs, and eventually humans.

Progress can seem slow when it comes to AGI software programming, but it only takes one innovative thinker to come up with an epiphany that could instantly accelerate the rate of advancement.

Current technology and computing power is insufficient for researchers to achieve artificial super intelligence today. However, the technology exists and the ideas are present. It will only be a matter of time before technology increases to such a degree to enable the necessary computing power needed for programmers and researchers to achieve their goal.

The Element of Innovation

Despite the huge amounts of money being invested and research going into the fields of robotic automation and artificial intelligence across a huge number of industries today, there is still the element of innovation to consider.

While the world watches for latest news on research and development reports released by major companies and forecasts for emerging technology being released in the near future, the very real possibility of an unknown enthusiast or researcher in a different field coming up with innovative breakthroughs in technology is also something to take into account.

Many of the world's most important technological inventions were created or pioneered by relatively unknown inventors or innovators. Yet they all play an integral role in the development of the technological revolution humanity is perched on the edge of today.

The world's first airplane was flown in 1903 by two unknown bicycle mechanics. 104 years later in 2007, the world's first computer-driven electronic automated airplane was flown using fly-by-wire technology.

An unknown Australian doctor and medical lecturer, Dr. Mark C Lidwill developed the first artificial pacemaker back in 1926. The original invention was used successfully in 1928 to revive a stillborn baby. An American physiologist, Albert Hyman, then released research in 1932 about his own experiments with an artificial pacemaker. The first external artificial pacemaker was then invented in 1951 by John Hopps, a Canadian electrical engineer who stumbled onto the medical use for the device while he was researching whether radio frequency heating could restore body temperature on a person suffering with hypothermia.

Almost a century later, cardiologists are able to implant intra-cardiac artificial pacemaker devices via a leg catheter, removing the need for invasive heart surgery. The self-contained miniature device is approximately the size of a pill and contains its own long-life power source.

The world's first bionic ear, or cochlear implant, brought the gift of sound back to the hearing impaired, thanks to a little-known Australian, Professor Graeme Clark back in the 1970s.

Wi-Fi technology was developed in 1992 by Australian John O'Sullivan, in conjunction with the CSIRO, using original research from the 1970s from radio astronomy. The technology is now used by more than a billion people all over the world to access the internet.

The world's first long-range guided rocket flight was launched in 1944 and was used by the military to launch ballistic missiles. By 2017, much of the world no longer views NASA's incredible guided rocket missions to outlying planets within our solar system and robotic missions to nearby planets as extraordinary.

The element of innovation could perhaps be the key to the next major advances in technology that once again change the way we think about robotics and AI in general.

Conclusion

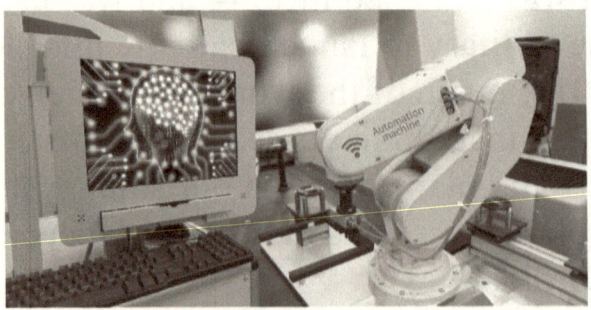

While it may still seem as though the concept of artificial super intelligence is a cool idea best left for the realms of science fiction, the reality remains that progress of robotics, automation and AI are advancing with each passing day.

As technology continues to progress, we learn to accept and adapt to the changes to how we do things and live our lives until those technological advances become commonplace.

Consider how long ago many of the original ideas and concepts for world-changing inventions were conceived and how long it took for them to become accepted into the mainstream. The basics of the technology needed were available on a limited basis, but only as technology increased and expanded did many of those inventions take off and become accepted as commonplace inclusions in our everyday lives.

The same principle is true with robotics and AI systems. It may have taken decades for the first AI systems to achieve low-level narrow-focus intelligence. But now that it's been achieved, the rate of acceleration has increased.

Reaching the next milestone of achieving artificial general intelligence systems may still be a few years away, but once it's been achieved it once again accelerates the rate of growth for moving on to the next phase.

Our entire society is at a major turning point in terms of how we think about work, career advancement, and income-earning potential. As technology continues to expand and grow, the sheer number of people who will be displaced and made redundant in their current occupations is increasingly likely.

The key to surviving and thriving in a future where robotics, automation and AI are likely to take over many jobs that no longer require human interaction could possibly lie in innovative thinking.

Take the time to understand the potential AI and robotic automation has to offer and then look for opportunities to create positive change. AI may be able to learn quickly, but it's certainly not flexible. After all, there are many new jobs out there that weren't available even five years ago.

Embrace the revolution and allow AI to complete the grunt work. Your focus should be on looking for new niches that open up as new technologies being to emerge.

Other Available Books:

- In The Pursuit of Wisdom: The Principal Thing

- **Investing in Gold and Silver Bullion - The Ultimate Safe Haven Investments**

- Nigerian Stock Market Investment: 2 Books with Bonus Content

- **The Dividend Millionaire: Investing for Income and Winning in the Stock Market**

- Economic Crisis: Surviving Global Currency Collapse - Safeguard Your Financial Future with Silver and Gold

- **Passionate about Stock Investing: The Quick Guide to Investing in the Stock Market**

- Guide to Investing in the Nigerian Stock Market

- Building Wealth with Dividend Stocks in the Nigerian Stock Market (Dividends - Stocks Secret Weapon)

- **Bitcoin and Digital Currency for Beginners: The Basic Little Guide**

- Child Millionaire: Stock Market Investing for Beginners - How to Build Wealth the Smart Way for Your Child

- **Christian Living: 2 Books with Bonus Content**

- Beginners Quick Guide to Passive Income: Learn Proven Ways to Earn Extra Income in the Cyber World

- **Taming the Tongue: The Power of Spoken Words**

- The Power of Positive Affirmations: Each Day a New Beginning

- The Real Estate Millionaire: Beginners Quick Start Guide to Investing In Properties and Learn How to Achieve Financial Freedom

- **Business: How to Quickly Make Real Money - Effective Methods to Make More Money: Easy and Proven Business Strategies for Beginners to Earn Even More Money in Your Spare Time**

- Money: Think Outside the Cube: 2-Book Money Making Boxed Set Bundle Strategies

- **Marketing: The Beginners Guide to Making Money Online with Social Media for Small Businesses**

- How to Effectively Lead and Win: The Proven Leadership Strategies and Techniques

If you would like to share this book with another person, please purchase an additional copy for each recipient. Thank you for your support and thanks for reading this book.

www.ingramcontent.com/pod-product-compliance
Lightning Source LLC
Chambersburg PA
CBHW021023180526
45163CB00005B/2095